Firefighter Tools

by Laura Hamilton Waxman

BUMBA BOOKS™

LERNER PUBLICATIONS ◆ MINNEAPOLIS

Note to Educators

Throughout this book, you'll find critical-thinking questions. These can be used to engage young readers in thinking critically about the topic and in using the text and photos to do so.

Lerner Publications Company
A division of Lerner Publishing Group, Inc.
241 First Avenue North
Minneapolis, MN 55401 USA

For reading levels and more information, look up this title at www.lernerbooks.com.

Main body text set in Helvetica Textbook Com Roman 23/49.
Typeface provided by Linotype AG.

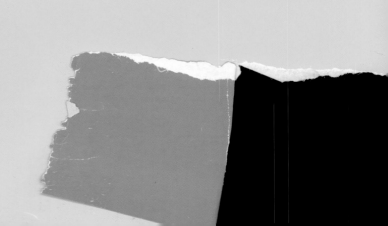

Library of Congress Cataloging-in-Publication Data

Names: Waxman, Laura Hamilton, author.
Title: Firefighter tools / Laura Hamilton Waxman.
Description: Minneapolis : Lerner Publications, [2020] | Series: Bumba books. Community helpers tools of the trade | Audience: Ages 4–7. | Audience: K to grade 3. | Includes index.
Identifiers: LCCN 2018043741 (print) | LCCN 2018046052 (ebook) | ISBN 9781541556492 (eb pdf) | ISBN 9781541555587 (lb : alk. paper)
Subjects: LCSH: Fire extinction—Equipment and supplies—Juvenile literature. | Fire fighters—Juvenile literature.
Classification: LCC TH9360 (ebook) | LCC TH9360 .W39 2020 (print) | DDC 628.9/25—dc23

LC record available at https://lccn.loc.gov/2018043741

Manufactured in the United States of America
1-46013-42930-11/28/2018

Table of
Contents

Firefighters

Firefighters rescue people and

places from fires.

Tools help them do their job.

A firefighter wears a suit, gloves,

and a helmet.

They help protect him from flames.

How else might a helmet keep firefighters safe?

Fires make smoke.

Firefighters may wear masks.

Masks bring them clean air.

A fire truck takes firefighters to a fire.

It has a siren and flashing lights.

Why do you think fire trucks have a siren and lights?

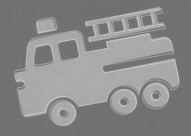

A fire truck carries hoses.

Firefighters use them to spray

water on fires.

Some fire trucks have a ladder.

Firefighters climb it to reach high places.

Sometimes firefighters use an ax.

It can break into burning buildings.

Firefighters carry a flashlight.

They use it to see in smoky places.

Firefighters use rope to climb from burning buildings.

It helps them escape fires quickly.

Firefighter Tools

fire truck

suit

hose

gloves

ladder

22

Picture Glossary

escape

to get away from

flames

glowing, hot parts of fire

rescue

to save someone or something

smoke

the cloud of gas that rises from a fire

23

Read More

Bellisario, Gina. *Firefighters in My Community*. Minneapolis: Lerner Publications, 2019.

Bowman, Chris. *Firefighters*. Minneapolis: Bellwether Media, 2018.

Reinke, Beth Bence. *Fire Trucks on the Go*. Minneapolis: Lerner Publications, 2018.

Index

Photo Credits

Image credits: bondvit/Shutterstock.com, pp. 5, 23; Jupiterimages/Getty Images, p. 6; Radius Images/Getty Images, pp. 9, 23; poco_bw/Getty Images, p. 10; Johan Mård/Folio/Getty Images, pp. 12–13; Stockbyte/Getty Images, p. 14; sirtravelalot/Shutterstock.com, p. 17; TatianaMironenko/Getty Images, pp. 18–19; ChiccoDodiFC/Shutterstock.com, pp. 21, 23; Tyler Olson/Shutterstock.com, p. 22; Phichai/Shutterstock.com, p. 22; Winai Tepsuttinun/Shutterstock.com, p. 22; USJ/Shutterstock.com, p. 22; Johner Images/Getty Images, p. 23.

Cover Images: Anest/Getty Images; Johan Mård/Folio/Getty Images; Ratthaphong Ekariyasap/Shutterstock.com.